BALANCING CHEMICAL EQUATIONS

THINGS YOU SHOULD KNOW
(QUESTIONS AND ANSWERS)

By Rumi Michael Leigh

Introduction

I would like to thank you for purchasing this book, *"Balancing chemical equations, things you should know (questions and answers)"*.

This book will help you understand, revise, and have a good general knowledge and understanding of balancing chemical equations.

I hope you enjoy it!

Table of Contents

Exercise 1: Balance the following chemical equations

Questions

    a) $Fe + HCl \rightarrow$

    b) $NH_3 + O_2 \rightarrow$

    c) $Na + H_2O \rightarrow$

    d) $Pb(NO_3)_2 + NaI \rightarrow$

    e) $C_2H_4 + O_2 \rightarrow$

    f) $Cu + AgNO_3 \rightarrow$

    g) $FeS + HNO_3 \rightarrow$

    h) $CaCO_3 + HCl \rightarrow$

    i) $Al + CuSO_4 \rightarrow$

Answers

    a) $2Fe + 6HCl \rightarrow 2FeCl_3 + 3H_2$

    b) $4NH_3 + 5O_2 \rightarrow 4NO + 6H_2O$

    c) $2Na + 2H_2O \rightarrow 2NaOH + H_2$

    d) $Pb(NO_3)_2 + 2NaI \rightarrow PbI_2 + 2NaNO_3$

    e) $C_2H_4 + 3O_2 \rightarrow 2CO_2 + 2H_2O$

    f) $2Cu + 2AgNO_3 \rightarrow 2Cu(NO_3)_2 + 2Ag$

    g) $2FeS + 6HNO_3 \rightarrow 2Fe(NO_3)_3 + 3H_2S$

    h) $CaCO_3 + 2HCl \rightarrow CaCl_2 + CO_2 + H_2O$

    i) $2Al + 3CuSO_4 \rightarrow Al_2(SO_4)_3 + 3Cu$

Exercise 2: Balance the following chemical equations

Questions

a) $C_4H_{10} + O_2 \rightarrow$

b) $Na_2SO_4 + BaCl_2 \rightarrow$

c) $Mg + HCl \rightarrow$

d) $Fe + H_2SO_4 \rightarrow$

e) $Cu + HNO_3 \rightarrow$

f) $Na + Cl_2 \rightarrow$

g) $K + H_2O \rightarrow$

h) $CH_4 + O_2 \rightarrow$

i) $C_6H_{12}O_6 + O_2 \rightarrow$

Answers

a) $C_4H_{10} + 13O_2 \rightarrow 8CO_2 + 10H_2O$

b) $Na_2SO_4 + BaCl_2 \rightarrow BaSO_4 + 2NaCl$

c) Balanced: $Mg + 2HCl \rightarrow MgCl_2 + H_2$

d) $Fe + H_2SO_4 \rightarrow FeSO_4 + H_2$

e) $3Cu + 8HNO_3 \rightarrow 3Cu(NO_3)_2 + 2NO + 4H_2O$

f) $2Na + Cl_2 \rightarrow 2NaCl$

g) $2K + 2H_2O \rightarrow 2KOH + H_2$

h) $CH_4 + 2O_2 \rightarrow CO_2 + 2H_2O$

i) $C_6H_{12}O_6 + 6O_2 \rightarrow 6CO_2 + 6H_2O$

Exercise 3: Balance the following chemical equations

Questions

a) $Al_2(SO_4)_3 + NaOH \rightarrow$

b) $Mg + H_2O \rightarrow$

c) $C_2H_6 + O_2 \rightarrow$

d) $Fe_2O_3 + CO \rightarrow$

e) $C_3H_8 + O_2 \rightarrow$

f) $CaO + H_2O \rightarrow$

g) $NaHCO_3 + HCl \rightarrow$

h) $AgNO_3 + Cu \rightarrow$

i) $Fe + CuSO_4 \rightarrow$

j) $K_2Cr_2O_7 + HCl \rightarrow$

Answers

a) $Al_2(SO_4)_3 + 6NaOH \rightarrow 3Na_2SO_4 + 2Al(OH)_3$

b) $Mg + 2H_2O \rightarrow Mg(OH)_2 + H_2$

c) $2C_2H_6 + 7O_2 \rightarrow 4CO_2 + 6H_2O$

d) $Fe_2O_3 + 3CO \rightarrow 2Fe + 3CO_2$

e) $C_3H_8 + 5O_2 \rightarrow 3CO_2 + 4H_2O$

f) $CaO + H_2O \rightarrow Ca(OH)_2$

g) $NaHCO_3 + HCl \rightarrow NaCl + CO_2 + H_2O$

h) $2AgNO_3 + Cu \rightarrow Cu(NO_3)_2 + 2Ag$

i) $Fe + CuSO_4 \rightarrow FeSO_4 + Cu$

j) $K_2Cr_2O_7 + 14HCl \rightarrow 2KCl + 2CrCl_3 + 7Cl_2 + 7H_2O$

Exercise 4: Balance the following chemical equations

Questions

a) $H_2SO_4 + NaOH \rightarrow$

b) $C_6H_6 + O_2 \rightarrow$

c) $CaCO_3 + HCl \rightarrow$

d) $NH_3 + O_2 \rightarrow$

e) $Al + HCl \rightarrow$

f) $FeS_2 + O_2 \rightarrow$

g) $C_8H_{18} + O_2 \rightarrow$

h) $Fe + H_2O \rightarrow$

i) $Ca(NO_3)_2 + NaOH \rightarrow$

j) $C_4H_6O_3 + O_2 \rightarrow$

Answers

a) $H_2SO_4 + 2NaOH \rightarrow Na_2SO_4 + 2H_2O$

b) $C_6H_6 + 15O_2 \rightarrow 6CO_2 + 3H_2O$

c) $CaCO_3 + 2HCl \rightarrow CaCl_2 + CO_2 + H_2O$

d) $4NH_3 + 5O_2 \rightarrow 4NO + 6H_2O$

e) $2Al + 6HCl \rightarrow 2AlCl_3 + 3H_2$

f) $4FeS_2 + 11O_2 \rightarrow 2Fe_2O_3 + 8SO_2$

g) $C_8H_{18} + 25O_2 \rightarrow 16CO_2 + 18H_2O$

h) $3Fe + 4H_2O \rightarrow Fe_3O_4 + 4H_2$

i) $Ca(NO_3)_2 + 2NaOH \rightarrow Ca(OH)_2 + 2NaNO_3$

j) $C_4H_6O_3 + 6O_2 \rightarrow 4CO_2 + 3H_2O$

Exercise 5: Balance the following chemical equations

Questions

a) $NH_3 + H_2SO_4 \rightarrow$

b) $HCl + CaCO_3 \rightarrow$

c) $C_2H_4 + O_2 \rightarrow$

d) $Na_2SO_4 + BaCl_2 \rightarrow$

e) $NH_3 + O_2 \rightarrow$

f) $NaClO + HCl \rightarrow$

g) $FeCl_3 + NaOH \rightarrow$

h) $C_6H_{12}O_6 + O_2 \rightarrow$

i) $NaOH + H_3PO_4 \rightarrow$

Answers

a) $2NH_3 + H_2SO_4 \rightarrow (NH_4)_2SO_4$

b) $2HCl + CaCO_3 \rightarrow CaCl_2 + CO_2 + H_2O$

c) $C_2H_4 + 3O_2 \rightarrow 2CO_2 + 2H_2O$

d) $Na_2SO_4 + BaCl_2 \rightarrow 2NaCl + BaSO_4$

e) $4NH_3 + 5O_2 \rightarrow 4NO + 6H_2O$

f) $NaClO + 2HCl \rightarrow NaCl + Cl_2 + H_2O$

g) $FeCl_3 + 3NaOH \rightarrow Fe(OH)_3 + 3NaCl$

h) $C_6H_{12}O_6 + 6O_2 \rightarrow 6CO_2 + 6H_2O$

i) $3NaOH + H_3PO_4 \rightarrow Na_3PO_4 + 3H_2O$

Exercise 6: Balance the following chemical equations

Questions

a) $Fe + HCl \rightarrow$

b) $K_2Cr_2O_7 + H_2SO_4 \rightarrow$

c) $Na_2CO_3 + HCl \rightarrow$

d) $Ca + H_2O \rightarrow$

e) $C_3H_8 + O_2 \rightarrow$

f) $Mg + HCl \rightarrow$

g) $NaOH + HCl \rightarrow$

h) $AgNO_3 + NaCl \rightarrow$

i) $NH_4NO_3 \rightarrow$

j) $Al + HNO_3 \rightarrow$

Answers

a) $Fe + 2HCl \rightarrow FeCl_2 + H_2$

b) $K_2Cr_2O_7 + 4H_2SO_4 \rightarrow K_2SO_4 + 2Cr_2(SO_4)_3 + 4H_2O + 3O_2$

c) $Na_2CO_3 + 2HCl \rightarrow 2NaCl + CO_2 + H_2O$

d) $Ca + 2H_2O \rightarrow Ca(OH)_2 + H_2$

e) $C_3H_8 + 5O_2 \rightarrow 3CO_2 + 4H_2O$

f) $Mg + 2HCl \rightarrow MgCl_2 + H_2$

g) $NaOH + HCl \rightarrow NaCl + H_2O$

h) $AgNO_3 + NaCl \rightarrow AgCl + NaNO_3$

i) $NH_4NO_3 \rightarrow N_2O + 2H_2O$

j) $2Al + 6HNO_3 \rightarrow 2Al(NO_3)_3 + 3NO + 3H_2O$

Exercise 7: Balance the following chemical equations

Questions

a) $FeCl_2 + NaOH \rightarrow$

b) $NaHCO_3 + HCl \rightarrow$

c) $H_2SO_4 + NaOH \rightarrow$

d) $Mg + O_2 \rightarrow$

e) $FeS_2 + O_2 \rightarrow$

f) $HNO_3 + Ba(OH)_2 \rightarrow$

g) $Na + Cl_2 \rightarrow$

h) $Al_2(SO_4)_3 + Ca(OH)_2 \rightarrow$

i) $CuO + H_2 \rightarrow$

Answers

a) $FeCl_2 + 2NaOH \rightarrow Fe(OH)_2 + 2NaCl$

b) $NaHCO_3 + HCl \rightarrow NaCl + CO_2 + H_2O$

c) $H_2SO_4 + 2NaOH \rightarrow Na_2SO_4 + 2H_2O$

d) $2Mg + O_2 \rightarrow 2MgO$

e) $4FeS_2 + 11O_2 \rightarrow 2Fe_2O_3 + 8SO_2$

f) $2HNO_3 + Ba(OH)_2 \rightarrow Ba(NO_3)_2 + 2H_2O$

g) $2Na + Cl_2 \rightarrow 2NaCl$

h) $Al_2(SO_4)_3 + 3Ca(OH)_2 \rightarrow 2Al(OH)_3 + 3CaSO_4$

i) $CuO + H_2 \rightarrow Cu + H_2O$

Exercise 8: Balance the following chemical equations

Questions

a) $NH_3 + O_2 \rightarrow$

b) $BaCl_2 + H_2SO_4 \rightarrow$

c) $H_2SO_4 + Al(OH)_3 \rightarrow$

d) $Pb(NO_3)_2 + Na_2SO_4 \rightarrow$

e) $C_6H_{12}O_6 + O_2 \rightarrow$

f) $K_2CrO_4 + HCl \rightarrow$

g) $FeS + HCl \rightarrow$

h) $KMnO_4 + HCl \rightarrow$

i) $C_2H_6O + O_2 \rightarrow$

j) $CaO + H_2O \rightarrow$

Answers

a) $4NH_3 + 5O_2 \rightarrow 4NO + 6H_2O$

b) $BaCl_2 + H_2SO_4 \rightarrow BaSO_4 + 2HCl$

c) $H_2SO_4 + 2Al(OH)_3 \rightarrow Al_2(SO_4)_3 + 6H_2O$

d) $Pb(NO_3)_2 + Na_2SO_4 \rightarrow PbSO_4 + 2NaNO_3$

e) $C_6H_{12}O_6 + 6O_2 \rightarrow 6CO_2 + 6H_2O$

f) $K_2CrO_4 + 2HCl \rightarrow 2KCl + CrO_2Cl_2 + H_2O$

g) $FeS + 2HCl \rightarrow FeCl_2 + H_2S$

h) $2KMnO_4 + 16HCl \rightarrow 2KCl + 2MnCl_2 + 5Cl_2 + 8H_2O$

i) $C_2H_6O + 3O_2 \rightarrow 2CO_2 + 3H_2O$

j) $CaO + H_2O \rightarrow Ca(OH)_2$

Exercise 9: Balance the following chemical equations

Questions

a) $CH_4 + O_2 \rightarrow$

b) $Fe + HCl \rightarrow$

c) $CuSO_4 + Zn \rightarrow$

d) $NH_3 + H_2O \rightarrow$

e) $Na_2CO_3 + HCl \rightarrow$

f) $Ca + HCl \rightarrow$

g) $Ba(OH)_2 + H_2SO_4 \rightarrow$

h) $C_4H_{10} + O_2 \rightarrow$

Answers

a) $CH_4 + 2O_2 \rightarrow CO_2 + 2H_2O$

b) $2Fe + 6HCl \rightarrow 2FeCl_3 + 3H_2$

c) $CuSO_4 + Zn \rightarrow ZnSO_4 + Cu$

d) $NH_3 + H_2O \rightarrow NH_4OH$

e) $Na_2CO_3 + 2HCl \rightarrow 2NaCl + CO_2 + H_2O$

f) $Ca + 2HCl \rightarrow CaCl_2 + H_2$

g) $Ba(OH)_2 + H_2SO_4 \rightarrow BaSO_4 + 2H_2O$

h) $C_4H_{10} + 13O_2 \rightarrow 8CO_2 + 10H_2O$

Exercise 10: Balance the following chemical equations

Questions

a) $NaOH + HCl \rightarrow$

b) $C2H5OH + O2 \rightarrow$

c) $Al + Fe2O3 \rightarrow$

d) $HNO3 + Cu \rightarrow$

e) $H2SO4 + NaOH \rightarrow$

f) $C6H6 + O2 \rightarrow$

g) $FeCl2 + NaOH \rightarrow$

h) $H3PO4 + Ca(OH)2 \rightarrow$

Answers

a) $NaOH + HCl \rightarrow NaCl + H2O$

b) $C2H5OH + 3O2 \rightarrow 2CO2 + 3H2O$

c) $2Al + Fe2O3 \rightarrow Al2O3 + 2Fe$

d) $4HNO3 + Cu \rightarrow Cu(NO3)2 + 2NO2 + 2H2O$

e) $H2SO4 + 2NaOH \rightarrow Na2SO4 + 2H2O$

f) $C6H6 + 15O2 \rightarrow 6CO2 + 3H2O$

g) $FeCl2 + 2NaOH \rightarrow Fe(OH)2 + 2NaCl$

h) $2H3PO4 + 3Ca(OH)2 \rightarrow Ca3(PO4)2 + 6H2O$

Exercise 11: Balance the following chemical equations

Questions

a) $Na_2S + HCl \rightarrow$

b) $CaCO_3 + HCl \rightarrow$

c) $Fe + H_2SO_4 \rightarrow$

d) $NaCl + AgNO_3 \rightarrow$

e) $C_2H_6 + O_2 \rightarrow$

f) $H_2S + O_2 \rightarrow$

g) $Mg + HCl \rightarrow$

h) $NH_3 + O_2 \rightarrow$

i) $Na + H_2O \rightarrow$

j) $Pb(NO_3)_2 + Na_2SO_4 \rightarrow$

Answers

a) $Na_2S + 2HCl \rightarrow 2NaCl + H_2S$

b) $CaCO_3 + 2HCl \rightarrow CaCl_2 + CO_2 + H_2O$

c) $Fe + H_2SO_4 \rightarrow FeSO_4 + H_2$

d) $NaCl + AgNO_3 \rightarrow AgCl + NaNO_3$

e) $2C_2H_6 + 7O_2 \rightarrow 4CO_2 + 6H_2O$

f) $2H_2S + 3O_2 \rightarrow 2SO_2 + 2H_2O$

g) $Mg + 2HCl \rightarrow MgCl_2 + H_2$

h) $4NH_3 + 5O_2 \rightarrow 4NO + 6H_2O$

i) $2Na + 2H_2O \rightarrow 2NaOH + H_2$

j) $Pb(NO_3)_2 + Na_2SO_4 \rightarrow PbSO_4 + 2NaNO_3$

Exercise 12: Balance the following chemical equations

Questions

a) $Al + H_2SO_4 \rightarrow$

b) $Cu + HNO_3 \rightarrow$

c) $C_3H_8 + O_2 \rightarrow$

d) $KMnO_4 + H_2C_2O_4 \rightarrow$

e) $FeCl_3 + NaOH \rightarrow$

f) $Ca + H_2O \rightarrow$

g) $H_2SO_4 + NH_4OH \rightarrow$

h) $H_2S + FeCl_3 \rightarrow$

i) $CaCl_2 + Na_2CO_3 \rightarrow$

Answers

a) $2Al + 3H_2SO_4 \rightarrow Al_2(SO_4)_3 + 3H_2$

b) $3Cu + 8HNO_3 \rightarrow 3Cu(NO_3)_2 + 2NO + 4H_2O$

c) $C_3H_8 + 5O_2 \rightarrow 3CO_2 + 4H_2O$

d) $2KMnO_4 + 5H_2C_2O_4 + 3H_2SO_4 \rightarrow K_2SO_4 + 2MnSO_4 + 10CO_2 + 8H_2O$

e) $FeCl_3 + 3NaOH \rightarrow Fe(OH)_3 + 3NaCl$

f) $Ca + 2H_2O \rightarrow Ca(OH)_2 + H_2$

g) $H_2SO_4 + 2NH_4OH \rightarrow (NH_4)_2SO_4 + 2H_2O$

h) $2H_2S + 2FeCl_3 \rightarrow 2FeS + 6HCl$

i) $CaCl_2 + Na_2CO_3 \rightarrow CaCO_3 + 2NaCl$

Exercise 13: Balance the following chemical equations

Questions

a) $NH_3 + H_2SO_4 \rightarrow$

b) $NaOH + H_3PO_4 \rightarrow$

c) $C_2H_5OH + O_2 \rightarrow$

d) $FeCl_2 + NaOH \rightarrow$

e) $H_2O_2 + KMnO_4 \rightarrow$

f) $CaO + H_2O \rightarrow$

g) $C_3H_8O + O_2 \rightarrow$

h) $HCl + NaOH \rightarrow$

i) $Fe + H_2O \rightarrow$

j) $C_6H_{12}O_6 + O_2 \rightarrow$

Answers

a) $2NH_3 + H_2SO_4 \rightarrow (NH_4)_2SO_4$

b) $3NaOH + H_3PO_4 \rightarrow Na_3PO_4 + 3H_2O$

c) $C_2H_5OH + 3O_2 \rightarrow 2CO_2 + 3H_2O$

d) $FeCl_2 + 2NaOH \rightarrow Fe(OH)_2 + 2NaCl$

e) $2H_2O_2 + 2KMnO_4 \rightarrow K_2O_2 + 2MnO_2 + 2O_2 + 2H_2O$

f) $CaO + H_2O \rightarrow Ca(OH)_2$

g) $C_3H_8O + 5O_2 \rightarrow 3CO_2 + 4H_2O$

h) $HCl + NaOH \rightarrow NaCl + H_2O$

i) $3Fe + 4H_2O \rightarrow Fe_3O_4 + 8H_2$

j) $C_6H_{12}O_6 + 6O_2 \rightarrow 6CO_2 + 6H_2O$

Exercise 14: Balance the following chemical equations

Questions

a) $Na + H_2O \rightarrow$

b) $AgNO_3 + HCl \rightarrow$

c) $Na_2O + H_2O \rightarrow$

d) $Fe + HCl \rightarrow$

e) $H_2SO_4 + NaOH \rightarrow$

f) $NH_3 + O_2 \rightarrow$

g) $C_4H_{10} + O_2 \rightarrow$

h) $Al + HCl \rightarrow$

Answers

a) $2Na + 2H_2O \rightarrow 2NaOH + H_2$

b) $AgNO_3 + HCl \rightarrow AgCl + HNO_3$

c) $Na_2O + H_2O \rightarrow 2NaOH$

d) $2Fe + 6HCl \rightarrow 2FeCl_3 + 3H_2$

e) $H_2SO_4 + 2NaOH \rightarrow Na_2SO_4 + 2H_2O$

f) $4NH_3 + 5O_2 \rightarrow 4NO + 6H_2O$

g) $C_4H_{10} + 13O_2 \rightarrow 8CO_2 + 10H_2O$

h) $2Al + 6HCl \rightarrow 2AlCl_3 + 3H_2$

Exercise 15: Balance the following chemical equations

Questions

a) $K_2Cr_2O_7 + HCl \rightarrow$

b) $C_2H_2 + O_2 \rightarrow$

c) $Fe_2O_3 + CO \rightarrow$

d) $H_2 + O_2 \rightarrow$

e) $Cu + HNO_3 \rightarrow$

f) $Mg + HCl \rightarrow$

g) $C_5H_{12} + O_2 \rightarrow$

h) $Fe + CuSO_4 \rightarrow$

i) $Na + Cl_2 \rightarrow$

j) $Pb(NO_3)_2 + KI \rightarrow$

Answers

a) $K_2Cr_2O_7 + 14HCl \rightarrow 2CrCl_3 + 3Cl_2 + 2KCl + 7H_2O$

b) $2C_2H_2 + 5O_2 \rightarrow 4CO_2 + 2H_2O$

c) $Fe_2O_3 + 3CO \rightarrow 2Fe + 3CO_2$

d) $2H_2 + O_2 \rightarrow 2H_2O$

e) $3Cu + 8HNO_3 \rightarrow 3Cu(NO_3)_2 + 2NO + 4H_2O$

f) $Mg + 2HCl \rightarrow MgCl_2 + H_2$

g) $C_5H_{12} + 8O_2 \rightarrow 5CO_2 + 6H_2O$

h) $Fe + CuSO_4 \rightarrow FeSO_4 + Cu$

i) $2Na + Cl_2 \rightarrow 2NaCl$

j) $Pb(NO_3)_2 + 2KI \rightarrow 2KNO_3 + PbI_2$

Exercise 16: Balance the following chemical equations

Questions

   a) $NH_3 + HCl \rightarrow$

   b) $C_6H_{12}O_6 + O_2 \rightarrow$

   c) $FeCl_2 + NaOH \rightarrow$

   d) $C_2H_5OH + O_2 \rightarrow$

   e) $Al + O_2 \rightarrow Al_2O_3$

   f) $Zn + H_2SO_4 \rightarrow$

   g) $H_2SO_4 + Ca(OH)_2 \rightarrow$

   h) $Pb + HNO_3 \rightarrow$

   i) $Na_2CO_3 + HCl \rightarrow$

Answers

   a) $NH_3 + HCl \rightarrow NH_4Cl$

   b) $C_6H_{12}O_6 + 6O_2 \rightarrow 6CO_2 + 6H_2O$

   c) $FeCl_2 + 2NaOH \rightarrow Fe(OH)_2 + 2NaCl$

   d) $2C_2H_5OH + 3O_2 \rightarrow 2CO_2 + 3H_2O$

   e) $4Al + 3O_2 \rightarrow 2Al_2O_3$

   f) $Zn + H_2SO_4 \rightarrow ZnSO_4 + H_2$

   g) $H_2SO_4 + Ca(OH)_2 \rightarrow CaSO_4 + 2H_2O$

   h) $3Pb + 8HNO_3 \rightarrow 3Pb(NO_3)_2 + 2NO + 4H_2O$

   i) $Na_2CO_3 + 2HCl \rightarrow 2NaCl + CO_2 + H_2O$

Exercise 17: Balance the following chemical equations

Questions

    a) $CuSO_4 + Fe \rightarrow$

    b) $NH_3 + O_2 \rightarrow$

    c) $AgNO_3 + NaCl \rightarrow$

    d) $C_2H_6 + O_2 \rightarrow$

    e) $HCl + NaOH \rightarrow$

    f) $Fe + HCl \rightarrow$

    g) $Al + HCl \rightarrow$

    h) $Mg + H_2SO_4 \rightarrow$

    i) $C_3H_8 + O_2 \rightarrow$

Answers

    a) $CuSO_4 + Fe \rightarrow FeSO_4 + Cu$

    b) $4NH_3 + 5O_2 \rightarrow 4NO + 6H_2O$

    c) $AgNO_3 + NaCl \rightarrow AgCl + NaNO_3$

    d) $2C_2H_6 + 7O_2 \rightarrow 4CO_2 + 6H_2O$

    e) $HCl + NaOH \rightarrow NaCl + H_2O$

    f) $Fe + 2HCl \rightarrow FeCl_2 + H_2$

    g) $2Al + 6HCl \rightarrow 2AlCl_3 + 3H_2$

    h) $Mg + H_2SO_4 \rightarrow MgSO_4 + H_2 + SO_2$

    i) $C_3H_8 + 5O_2 \rightarrow 3CO_2 + 4H_2O$

Exercise 18: Balance the following chemical equations

Questions

a) $CaO + H_2O \rightarrow$

b) $C_8H_{18} + O_2 \rightarrow$

c) $Na + H_2O \rightarrow$

d) $Fe + H_2O \rightarrow$

e) $NaHCO_3 + HCl \rightarrow$

f) $Pb(NO_3)_2 + KI \rightarrow$

g) $C_2H_4 + O_2 \rightarrow$

h) $Mg + HCl \rightarrow$

i) $NH_3 + O_2 \rightarrow$

Answers

a) $CaO + H_2O \rightarrow Ca(OH)_2$

b) $2C_8H_{18} + 25O_2 \rightarrow 16CO_2 + 18H_2O$

c) $2Na + 2H_2O \rightarrow 2NaOH + H_2$

d) $3Fe + 4H_2O \rightarrow Fe_3O_4 + 4H_2$

e) $NaHCO_3 + HCl \rightarrow NaCl + CO_2 + H_2O$

f) $Pb(NO_3)_2 + 2KI \rightarrow PbI_2 + 2KNO_3$

g) $C_2H_4 + 3O_2 \rightarrow 2CO_2 + 2H_2O$

h) $Mg + 2HCl \rightarrow MgCl_2 + H_2$

i) $4NH_3 + 5O_2 \rightarrow 4NO + 6H_2O$

Exercise 19: Balance the following chemical equations

Questions

a) $C_6H_{12}O_6 + O_2 \rightarrow$

b) $CaCO_3 + HCl \rightarrow$

c) $H_2SO_4 + NaOH \rightarrow$

d) $FeCl_2 + NaOH \rightarrow$

e) $C_8H_{18} + O_2 \rightarrow$

f) $Na_2CO_3 + HCl \rightarrow$

g) $NH_3 + HCl \rightarrow$

h) $C_6H_6 + O_2 \rightarrow$

i) $H_3PO_4 + Ca(OH)_2 \rightarrow$

Answers

a) $C_6H_{12}O_6 + 6O_2 \rightarrow 6CO_2 + 6H_2O$

b) $CaCO_3 + 2HCl \rightarrow CaCl_2 + CO_2 + H_2O$

c) $H_2SO_4 + 2NaOH \rightarrow Na_2SO_4 + 2H_2O$

d) $FeCl_2 + 2NaOH \rightarrow Fe(OH)_2 + 2NaCl$

e) $2C_8H_{18} + 25O_2 \rightarrow 16CO_2 + 18H_2O$

f) $Na_2CO_3 + 2HCl \rightarrow 2NaCl + CO_2 + H_2O$

g) $NH_3 + HCl \rightarrow NH_4Cl$

h) $C_6H_6 + 15O_2 \rightarrow 6CO_2 + 3H_2O$

i) $2H_3PO_4 + 3Ca(OH)_2 \rightarrow Ca_3(PO_4)_2 + 6H_2O$

Exercise 20: Balance the following chemical equations

Questions

a) $Fe + HCl \rightarrow$

b) $Mg + HNO_3$

c) $H_2 + O_2 \rightarrow$

d) $K_2Cr_2O_7 + FeSO_4 + H_2SO_4 \rightarrow$

e) $FeCl_3 + NaOH \rightarrow$

f) $C_2H_6 + O_2 \rightarrow$

g) $BaCl_2 + Na_2SO_4 \rightarrow$

h) $H_2SO_4 + Mg \rightarrow$

i) $K_2CrO_4 + HCl \rightarrow$

Answers

a) $2Fe + 6HCl \rightarrow 2FeCl_3 + 3H_2$

b) $Mg + 4HNO_3 \rightarrow Mg(NO_3)_2 + 2NO_2 + 2H_2O$

c) $2H_2 + O_2 \rightarrow 2H_2O$

d) $K_2Cr_2O_7 + 4FeSO_4 + 8H_2SO_4 \rightarrow K_2SO_4 + 4Fe_2(SO_4)_3 + Cr_2(SO_4)_3 + 8H_2O$

e) $FeCl_3 + 3NaOH \rightarrow Fe(OH)_3 + 3NaCl$

f) $2C_2H_6 + 7O_2 \rightarrow 4CO_2 + 6H_2O$

g) $BaCl_2 + Na_2SO_4 \rightarrow BaSO_4 + 2NaCl$

h) $H_2SO_4 + Mg \rightarrow MgSO_4 + H_2$

i) $K_2CrO_4 + 2HCl \rightarrow 2KCl + CrCl_3 + H_2O$

Exercise 21: Balance the following chemical equations

Questions

a) $NH_3 + O_2 \rightarrow$
b) $CH_4 + O_2 \rightarrow$
c) $Cu + HNO_3 \rightarrow$
d) $Pb(NO_3)_2 + KI \rightarrow$
e) $H_2SO_4 + NaOH \rightarrow$
f) $Al + H_2SO_4 \rightarrow$
g) $CaCO_3 + HCl \rightarrow$
h) $C_2H_5OH + O_2 \rightarrow$
i) $Na_2CO_3 + HCl \rightarrow$

Answers

a) $4NH_3 + 5O_2 \rightarrow 4NO + 6H_2O$
b) $CH_4 + 2O_2 \rightarrow CO_2 + 2H_2O$
c) $3Cu + 8HNO_3 \rightarrow 3Cu(NO_3)_2 + 2NO + 4H_2O$
d) $Pb(NO_3)_2 + 2KI \rightarrow PbI_2 + 2KNO_3$
e) $H_2SO_4 + 2NaOH \rightarrow Na_2SO_4 + 2H_2O$
f) $2Al + 3H_2SO_4 \rightarrow Al_2(SO_4)_3 + 3H_2$
g) $CaCO_3 + 2HCl \rightarrow CaCl_2 + CO_2 + H_2O$
h) $C_2H_5OH + 3O_2 \rightarrow 2CO_2 + 3H_2O$
i) $Na_2CO_3 + 2HCl \rightarrow 2NaCl + CO_2 + H_2O$

Exercise 22: Balance the following chemical equations

Questions

a) $Fe + HCl \rightarrow$

b) $H2 + O2 \rightarrow$

c) $NH3 + HCl \rightarrow$

d) $FeS2 + O2 \rightarrow$

e) $Al2O3 + H2SO4 \rightarrow$

f) $Mg + H2SO4 \rightarrow$

g) $HCl + NaOH \rightarrow$

h) $C6H12O6 + O2 \rightarrow$

i) $Pb + HNO3 \rightarrow$

Answers

a) $2Fe + 6HCl \rightarrow 2FeCl3 + 3H2$

b) $2H2 + O2 \rightarrow 2H2O$

c) $NH3 + HCl \rightarrow NH4Cl$

d) $4FeS2 + 11O2 \rightarrow 2Fe2O3 + 8SO2$

e) $Al2O3 + 3H2SO4 \rightarrow Al2(SO4)3 + 3H2O$

f) $Mg + H2SO4 \rightarrow MgSO4 + H2$

g) $HCl + NaOH \rightarrow NaCl + H2O$

h) $C6H12O6 + 6O2 \rightarrow 6CO2 + 6H2O$

i) $3Pb + 8HNO3 \rightarrow 3Pb(NO3)2 + 2NO + 4H2O$

Exercise 23: Balance the following chemical equations

Questions

a) $Ca(OH)_2 + H_3PO_4 \rightarrow$

b) $C_2H_5OH + O_2 \rightarrow$

c) $Fe + CuSO_4 \rightarrow$

d) $Na_2CO_3 + HCl \rightarrow$

e) $C_8H_{18} + O_2 \rightarrow$

f) $MgO + HNO_3 \rightarrow$

g) $Cu + HNO_3 \rightarrow$

h) $H_2S + KMnO_4 \rightarrow$

i) $NH_3 + O_2 \rightarrow$

Answers

a) $3Ca(OH)_2 + 2H_3PO_4 \rightarrow Ca_3(PO_4)_2 + 6H_2O$

b) $C_2H_5OH + 3O_2 \rightarrow 2CO_2 + 3H_2O$

c) $Fe + CuSO_4 \rightarrow FeSO_4 + Cu$

d) $Na_2CO_3 + 2HCl \rightarrow 2NaCl + CO_2 + H_2O$

e) $C_8H_{18} + 25O_2 \rightarrow 16CO_2 + 18H_2O$

f) $MgO + 2HNO_3 \rightarrow Mg(NO_3)_2 + H_2O$

g) $3Cu + 8HNO_3 \rightarrow 3Cu(NO_3)_2 + 2NO + 4H_2O$

h) $5H_2S + 8KMnO_4 \rightarrow 5MnS + K_2SO_4 + 8H_2O + 8O_2$

i) $4NH_3 + 5O_2 \rightarrow 4NO + 6H_2O$

Exercise 24: Balance the following chemical equations

Questions

a) $Na_2SO_4 + Pb(NO_3)_2 \rightarrow$

b) $Mg + H_2SO_4 \rightarrow$

c) $FeS_2 + O_2 \rightarrow$

d) $H_2SO_4 + NaOH \rightarrow$

e) $C_6H_{12}O_6 + O_2 \rightarrow$

f) $Zn + HCl \rightarrow$

g) $AlCl_3 + NaOH \rightarrow$

h) $C_3H_8 + O_2 \rightarrow$

i) $Fe + HCl \rightarrow$

Answers

a) $Na_2SO_4 + Pb(NO_3)_2 \rightarrow PbSO_4 + 2NaNO_3$

b) $Mg + H_2SO_4 \rightarrow MgSO_4 + H_2$

c) $4FeS_2 + 11O_2 \rightarrow 2Fe_2O_3 + 8SO_2$

d) $H_2SO_4 + 2NaOH \rightarrow Na_2SO_4 + 2H_2O$

e) $C_6H_{12}O_6 + 6O_2 \rightarrow 6CO_2 + 6H_2O$

f) $Zn + 2HCl \rightarrow ZnCl_2 + H_2$

g) $AlCl_3 + 3NaOH \rightarrow Al(OH)_3 + 3NaCl$

h) $C_3H_8 + 5O_2 \rightarrow 3CO_2 + 4H_2O$

i) $2Fe + 6HCl \rightarrow 2FeCl_3 + 3H_2$

Exercise 25: Balance the following chemical equations

Questions

a) $NH_3 + O_2 \rightarrow$

b) $C_2H_6O + O_2 \rightarrow$

c) $Ca(OH)_2 + H_3PO_4 \rightarrow$

d) $Na + H_2O \rightarrow$

e) $FeCl_2 + NaOH \rightarrow$

f) $C_8H_{18} + O_2 \rightarrow$

g) $AgNO_3 + NaCl \rightarrow$

h) $Pb(NO_3)_2 + KI \rightarrow$

i) $H_2SO_4 + Fe \rightarrow$

j) $HNO_3 + Ca(OH)_2 \rightarrow$

Answers

a) $4NH_3 + 5O_2 \rightarrow 4NO + 6H_2O$

b) $C_2H_6O + 3O_2 \rightarrow 2CO_2 + 3H_2O$

c) $3Ca(OH)_2 + 2H_3PO_4 \rightarrow Ca_3(PO_4)_2 + 6H_2O$

d) $2Na + 2H_2O \rightarrow 2NaOH + H_2$

e) $FeCl_2 + 2NaOH \rightarrow Fe(OH)_2 + 2NaCl$

f) $C_8H_{18} + 25O_2 \rightarrow 16CO_2 + 18H_2O$

g) $AgNO_3 + NaCl \rightarrow AgCl + NaNO_3$

h) $Pb(NO_3)_2 + 2KI \rightarrow PbI_2 + 2KNO_3$

i) $H_2SO_4 + Fe \rightarrow FeSO_4 + H_2$

j) $2HNO_3 + Ca(OH)_2 \rightarrow Ca(NO_3)_2 + 2H_2O$

Exercise 26: Balance the following chemical equations

Questions

a) $NaHCO_3 + HCl \rightarrow$

b) $Pb + HNO_3 \rightarrow$

c) $Mg + HCl \rightarrow$

d) $CaCO_3 + HCl \rightarrow$

e) $K_2Cr_2O_7 + HCl \rightarrow$

f) $C_6H_{12}O_6 + O_2 \rightarrow$

g) $Fe + H_2SO_4 \rightarrow$

h) $AlCl_3 + Ca(OH)_2 \rightarrow$

Answers

a) $NaHCO_3 + HCl \rightarrow NaCl + CO_2 + H_2O$

b) $3Pb + 8HNO_3 \rightarrow 3Pb(NO_3)_2 + 2NO + 4H_2O$

c) $Mg + 2HCl \rightarrow MgCl_2 + H_2$

d) $CaCO_3 + 2HCl \rightarrow CaCl_2 + CO_2 + H_2O$

e) $K_2Cr_2O_7 + 14HCl \rightarrow 2CrCl_3 + 3Cl_2 + 7H_2O + 2KCl$

f) $C_6H_{12}O_6 + 6O_2 \rightarrow 6CO_2 + 6H_2O$

g) $Fe + H_2SO_4 \rightarrow FeSO_4 + H_2$

h) $2AlCl_3 + 3Ca(OH)_2 \rightarrow 2Al(OH)_3 + 3CaCl_2$

Exercise 27: Balance the following chemical equations

Questions

a) $FeCl_2 + KMnO_4 + HCl \rightarrow$

b) $Na_2O_2 + H_2SO_4 \rightarrow$

c) $AgNO_3 + Cu \rightarrow$

d) $Pb(NO_3)_2 + K_2CrO_4 \rightarrow$

e) $NH_4Cl + Ca(OH)_2 \rightarrow$

f) $NaOH + H_2SO_4 \rightarrow$

g) $C_2H_5OH + O_2 \rightarrow$

h) $NH_3 + O_2 \rightarrow$

i) $H_2 + Cl_2 \rightarrow$

j) $Cu + HNO_3 \rightarrow$

Answers

a) $5FeCl_2 + 2KMnO_4 + 8HCl \rightarrow 5FeCl_3 + 2KCl + 2MnCl_2 + 4H_2O$

b) $Na_2O_2 + H_2SO_4 \rightarrow Na_2SO_4 + H_2O_2$

c) $2AgNO_3 + Cu \rightarrow Cu(NO_3)_2 + 2Ag$

d) $Pb(NO_3)_2 + K_2CrO_4 \rightarrow PbCrO_4 + 2KNO_3$

e) $2NH_4Cl + Ca(OH)_2 \rightarrow 2NH_3 + CaCl_2 + 2H_2O$

f) $2NaOH + H_2SO_4 \rightarrow Na_2SO_4 + 2H_2O$

g) $C_2H_5OH + 3O_2 \rightarrow 2CO_2 + 3H_2O$

h) $4NH_3 + 5O_2 \rightarrow 4NO + 6H_2O$

i) $H_2 + Cl_2 \rightarrow 2HCl$

j) $3Cu + 8HNO_3 \rightarrow 3Cu(NO_3)_2 + 2NO + 4H_2O$

Exercise 28: Balance the following chemical equations

Questions

a) $K_4Fe(CN)_6 + KMnO_4 + H_2SO_4 \rightarrow$

b) $C_3H_8 + O_2 \rightarrow$

c) $CaCO_3 + HNO_3 \rightarrow$

d) $P_4 + O_2 \rightarrow$

e) $Al + HCl \rightarrow$

f) $NH_3 + HCl \rightarrow$

g) $Fe + H_2SO_4 \rightarrow$

h) $Mg + HCl \rightarrow$

i) $C_6H_{12}O_6 + O_2 \rightarrow$

Answers

a) $3K_4Fe(CN)_6 + 2KMnO_4 + 18H_2SO_4 \rightarrow 3KHSO_4 + 2Fe_2(SO_4)_3 + 2MnSO_4 + 5HNO_3 + 6CO_2 + 18H_2O$

b) $C_3H_8 + 5O_2 \rightarrow 3CO_2 + 4H_2O$

c) $CaCO_3 + 2HNO_3 \rightarrow Ca(NO_3)_2 + CO_2 + H_2O$

d) $P_4 + 5O_2 \rightarrow 2P_2O_5$

e) $2Al + 6HCl \rightarrow 2AlCl_3 + 3H_2$

f) $NH_3 + HCl \rightarrow NH_4Cl$

g) $Fe + H_2SO_4 \rightarrow FeSO_4 + H_2O + SO_2$

h) $Mg + 2HCl \rightarrow MgCl_2 + H_2$

i) $C_6H_{12}O_6 + 6O_2 \rightarrow 6CO_2 + 6H_2O$

Exercise 29: Balance the following chemical equations

Questions

a) $Fe_2O_3 + CO \rightarrow$
b) $Na + H_2O \rightarrow$
c) $C_2H_5OH + O_2 \rightarrow$
d) $HCl + Ca(OH)_2 \rightarrow$
e) $Mg + H_2SO_4 \rightarrow$
f) $Pb(NO_3)_2 + KI \rightarrow$
g) $C_2H_2 + O_2 \rightarrow$
h) $H_2SO_4 + KOH \rightarrow$

Answers

a) $Fe_2O_3 + 3CO \rightarrow 2Fe + 3CO_2$
b) $2Na + 2H_2O \rightarrow 2NaOH + H_2$
c) $C_2H_5OH + 3O_2 \rightarrow 2CO_2 + 3H_2O$
d) $2HCl + Ca(OH)_2 \rightarrow CaCl_2 + 2H_2O$
e) $Mg + H_2SO_4 \rightarrow MgSO_4 + H_2$
f) $Pb(NO_3)_2 + 2KI \rightarrow 2KNO_3 + PbI_2$
g) $2C_2H_2 + 5O_2 \rightarrow 4CO_2 + 2H_2O$
h) $H_2SO_4 + 2KOH \rightarrow K_2SO_4 + 2H_2O$

Exercise 30: Balance the following chemical equations

Questions

a) $Mg + HCl \rightarrow$

b) $Cu + HNO_3 \rightarrow$

c) $NaHCO_3 + HCl \rightarrow$

d) $NH_3 + O_2 \rightarrow$

e) $Al + H_2SO_4 \rightarrow$

f) $C_6H_{12}O_6 + O_2 \rightarrow$

g) $CaCO_3 + HCl \rightarrow$

h) $Fe + H_2SO_4 \rightarrow$

i) $K_2Cr_2O_7 + H_2SO_4 + NaSO_3 \rightarrow$

j) $Pb(NO_3)_2 + NaI \rightarrow$

Answers

a) $Mg + 2HCl \rightarrow MgCl_2 + H_2$

b) $3Cu + 8HNO_3 \rightarrow 3Cu(NO_3)_2 + 2NO + 4H_2O$

c) $NaHCO_3 + HCl \rightarrow NaCl + CO_2 + H_2O$

d) $4NH_3 + 5O_2 \rightarrow 4NO + 6H_2O$

e) $2Al + 3H_2SO_4 \rightarrow Al_2(SO_4)_3 + 3H_2$

f) $C_6H_{12}O_6 + 6O_2 \rightarrow 6CO_2 + 6H_2O$

g) $CaCO_3 + 2HCl \rightarrow CaCl_2 + CO_2 + H_2O$

h) $Fe + H_2SO_4 \rightarrow FeSO_4 + H_2$

i) $K_2Cr_2O_7 + 4H_2SO_4 + 3Na_2SO_3 \rightarrow Cr_2(SO_4)_3 + 3H_2O + 3Na_2SO_4 + K_2SO_4$

j) $Pb(NO_3)_2 + 2NaI \rightarrow 2NaNO_3 + PbI_2$

Exercise 31: Balance the following chemical equations

Questions

a) $NH_4Cl + Ca(OH)_2 \rightarrow$

b) $NaOH + H_3PO_4 \rightarrow$

c) $Ca + H_2O \rightarrow$

d) $C_3H_8 + O_2 \rightarrow$

e) $Na + H_2O \rightarrow$

f) $Fe_2O_3 + H_2 \rightarrow$

g) $Al_2O_3 + HCl \rightarrow$

h) $Mg + H_2SO_4 \rightarrow$

i) $Zn + HCl \rightarrow$

Answers

a) $2NH_4Cl + Ca(OH)_2 \rightarrow 2NH_3 + 2H_2O + CaCl_2$

b) $3NaOH + H_3PO_4 \rightarrow Na_3PO_4 + 3H_2O$

c) $Ca + 2H_2O \rightarrow Ca(OH)_2 + H_2$

d) $C_3H_8 + 5O_2 \rightarrow 3CO_2 + 4H_2O$

e) $2Na + 2H_2O \rightarrow 2NaOH + H_2$

f) $Fe_2O_3 + 3H_2 \rightarrow 2Fe + 3H_2O$

g) $Al_2O_3 + 6HCl \rightarrow 2AlCl_3 + 3H_2O$

h) $Mg + H_2SO_4 \rightarrow MgSO_4 + H_2$

i) $Zn + 2HCl \rightarrow ZnCl_2 + H_2$

Exercise 32: Balance the following chemical equations

Questions

a) $Fe + HNO_3 \rightarrow$

b) $C_2H_5OH + O_2 \rightarrow$

c) $H_2SO_4 + NaOH \rightarrow$

d) $Na_2CO_3 + HCl \rightarrow$

e) $Na_2S_2O_3 + Cl_2 \rightarrow$

f) $HCl + NaHCO_3 \rightarrow$

g) $C_4H_{10} + O_2 \rightarrow$

h) $Al + H_2SO_4 \rightarrow$

i) $CaCO_3 + HNO_3 \rightarrow$

j) $C_6H_6 + O_2 \rightarrow$

Answers

a) $4Fe + 10HNO_3 \rightarrow 4Fe(NO_3)_3 + 3NO + 5H_2O$

b) $C_2H_5OH + 3O_2 \rightarrow 2CO_2 + 3H_2O$

c) $H_2SO_4 + 2NaOH \rightarrow Na_2SO_4 + 2H_2O$

d) $Na_2CO_3 + 2HCl \rightarrow 2NaCl + CO_2 + H_2O$

e) $Na_2S_2O_3 + 2Cl_2 \rightarrow 2NaCl + Na_2S_4O_6$

f) $HCl + NaHCO_3 \rightarrow NaCl + CO_2 + H_2O$

g) $C_4H_{10} + 13/2O_2 \rightarrow 4CO_2 + 5H_2O$

h) $2Al + 3H_2SO_4 \rightarrow Al_2(SO_4)_3 + 3H_2$

i) $CaCO_3 + 2HNO_3 \rightarrow Ca(NO_3)_2 + CO_2 + H_2O$

j) $C_6H_6 + 15/2O_2 \rightarrow 6CO_2 + 3H_2O$

Exercise 33: Balance the following chemical equations

Questions

a) $C_6H_{12}O_6 + O_2 \rightarrow$

b) $AgNO_3 + NaCl \rightarrow$

c) $K_2Cr_2O_7 + H_2SO_4 \rightarrow$

d) $Na_2O + H_2O \rightarrow$

e) $C_{12}H_{22}O_{11} + H_2O \rightarrow$

f) $NaOH + HCl \rightarrow$

g) $Fe_2O_3 + CO \rightarrow$

h) $NH_3 + O_2 \rightarrow$

i) $CH_4 + O_2 \rightarrow$

Answers

a) $C_6H_{12}O_6 + 6O_2 \rightarrow 6CO_2 + 6H_2O$

b) $AgNO_3 + NaCl \rightarrow AgCl + NaNO_3$

c) $K_2Cr_2O_7 + 4H_2SO_4 \rightarrow K_2SO_4 + Cr_2(SO_4)_3 + 4H_2O + 3O_2$

d) $Na_2O + H_2O \rightarrow 2NaOH$

e) $C_{12}H_{22}O_{11} + 12H_2O \rightarrow 6C_6H_{12}O_6$

f) $NaOH + HCl \rightarrow NaCl + H_2O$

g) $Fe_2O_3 + 3CO \rightarrow 2Fe + 3CO_2$

h) $4NH_3 + 5O_2 \rightarrow 4NO + 6H_2O$

i) $CH_4 + 2O_2 \rightarrow CO_2 + 2H_2O$

Exercise 34: Balance the following chemical equations

Questions

a) $CaO + H_2O \rightarrow$

b) $FeCl_2 + K_2Cr_2O_7 + HCl \rightarrow$

c) $Na + H_2O \rightarrow$

d) $C_3H_8 + O_2 \rightarrow$

e) $NH_3 + HCl \rightarrow$

f) $Mg + HCl \rightarrow$

g) $Na_2CO_3 + HCl \rightarrow$

h) $NaHCO_3 \rightarrow$

i) $CaCO_3 + HCl \rightarrow$

Answers

a) $CaO + H_2O \rightarrow Ca(OH)_2$

b) $6FeCl_2 + K_2Cr_2O_7 + 14HCl \rightarrow 6FeCl_3 + 2CrCl_3 + 2KCl + 7H_2O$

c) $2Na + 2H_2O \rightarrow 2NaOH + H_2$

d) $C_3H_8 + 5O_2 \rightarrow 3CO_2 + 4H_2O$

e) $NH_3 + HCl \rightarrow NH_4Cl$

f) $Mg + 2HCl \rightarrow MgCl_2 + H_2$

g) $Na_2CO_3 + 2HCl \rightarrow 2NaCl + H_2O + CO_2$

h) $2NaHCO_3 \rightarrow Na_2CO_3 + CO_2 + H_2O$

i) $CaCO_3 + 2HCl \rightarrow CaCl_2 + CO_2 + H_2O$

Conclusion

Thank you once again for purchasing this book. I hope it has helped you in your journey to understand the basics of balancing chemical equations.

Please, if you learnt something from this book, I would like you to leave a review. It'd be appreciated.

Thank you.